2nd Grade Science Volume 2

© 2013 Todd Deluca
OnBoard Academics, Inc
Newburyport, MA 01950

800-596-3175
www.onboardacademics.com

Table of Contents

Soil

Where does soil come from?
Soil is formed when rock at the earths surface is broken down into tiny grains from weathering and erosion. Weathering and erosion describe the process by which forces such as freezing, wind and rain wear and break down rock into tiny particles sometimes over many many millions of years. The tiny particles of rock are often carried long distances from the parent rock usually by forces such as winds and rivers.

Fertile soil is the name we give to soil in which plants will grow. Fertile soil contains humus, waste from animals as well as dead animals and plants that become part of the soil. Humus adds nutrients to the soil which is the food source to the many millions of microscopic organisms found in solid that help to keep solid healthy. The other elements in soil are water and air. Air and water occupy the gaps between soil particles.

The diagram that shows the composition of a typical soil.

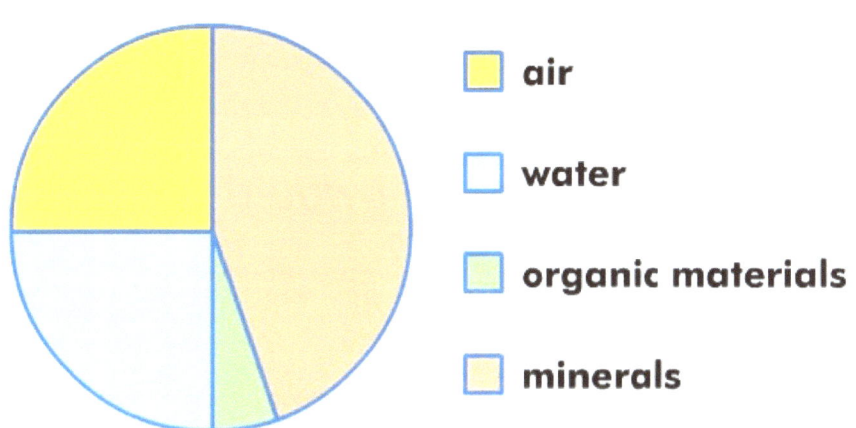

air

water

organic materials

minerals

There are many thousands of soils ranging in depths from a few centimeters to tens of meters.   We classify them as sandy, clay an loam.

Sandy soil is mostly sand and a little bit of clay. This soil would feel gritty in your hands. Sandy soils have very little humus and don't hold water well so they aren't' very fertile.

Clay soils are made up mostly of clay with some sand. Clay would feel smooth and sticky in your hands. Clay holds water well but hardens like a rock when dry. This makes growing plants in clay challenging.

Loam is a mixture of sand, clay and humus. It looks very dark in color and can hold enough water and is full of organic matter making it an excellent soil in which to grow plants.

Label the typical composition of soil.

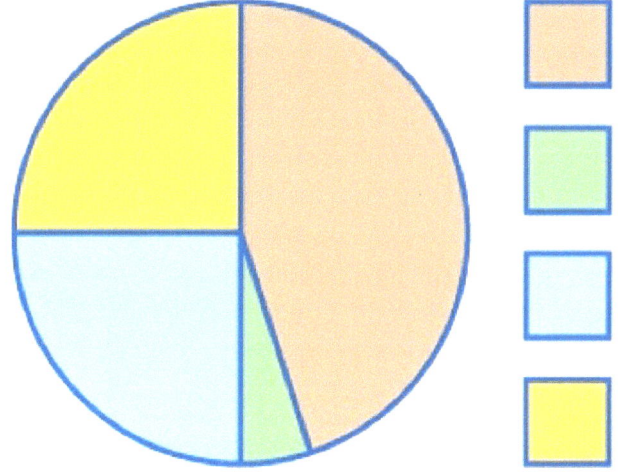

humus   **water**   **minerals**   **air**

Examine the color of the soil and the condition of the plant and then label each pot with the type of soil inside.

| Loamy | Sandy | Clay |

Plants depend upon minerals from the soil, like nitrogen and phosphorus, to grow and to make their own food. Some crops like corn use a lot of nitrogen, while crops like alfalfa add nitrogen to the soil. Farmers rotate their crops to keep the soil healthy and enriched. Farmers also use fertilizers to replace lost minerals.

Soil Erosion

Although weathering and erosion are key forces in creating soil, soil is also subject to erosion. Rain and running water are the biggest culprits as they loosen sol and then it is washed away.

Humans also cause soil erosion. Plowing breaks up and loosens soil. Clearing trees and shrubs increases erosion since roots help to bind soil an keep it in place.

To reduce erosion, people use a number of techniques such as planting trees and shrubs, contour plowing which means to plow along the natural surface instead of straight lines and terrace plowing which involves building reinforce flat ledges on hillsides and growing trees and plants that shelter crops and prevent solid from being blown away.

Soil Quiz

1. There are many _____ of different types of soil.
   - a. hundreds
   - b. thousands
   - c. millions

2. Sandy should contains very little _____ and does not hold water well.
   - a. clay
   - b. humus
   - c. loam
   - d. rocks

3. Clay is an excellent soil in which to grow plants. True or false?

4. Farmers use _____ to replace lost minerals in soil.
   - a. vitamins
   - b. nitrogen
   - c. fertilizers

5. Soil is formed when rocks on the Earth's surface are broke down as a result of weathering and _____.
   - a. climate
   - b. plowing
   - c. erosion
   - d. floods

Plant Life Cycle

From all of these plants, only two did not start off as seeds.

moss	grass	rose	pineapple	broccoli
tomato	carrot	oak	fir	fern

moss

fern

Most plants start their life cycle as a seed. To understand the life-cycle of plants, you first need to understand how seeds work.

Parts of a Seed.

Most plants begin their life as a seed.

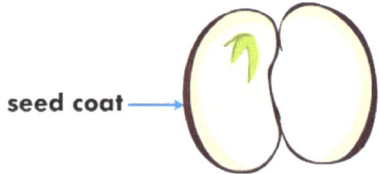

The outside of a seed is the coat. It protects the the inside of the seed like a coat of armor.

Inside the seed there are two parts.

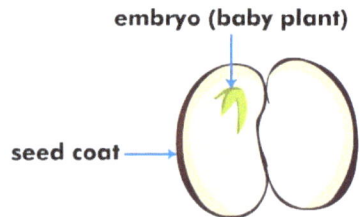

The first part is the baby plant also called the embryo. The embryo is the part of the plant that will grow to form an adult plant.

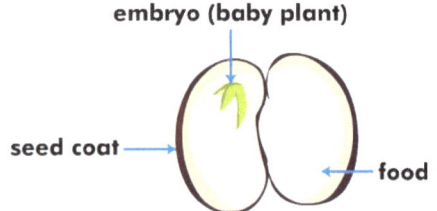

The second part is the food. This part supplies the food to the baby plant since it can't make its own food. The seed needs to provide the food until the plant is able to grow leaves and create its own food.

Label the description with the seed part.

I'm the baby plant.	
I protect the embryo.	
I'm an energy source for the embryo.	
I'll grow into an adult plant.	

E — embryo _____

F — food _____

S — seed coat _____

A seed grows into a plant.

If there is a right balance between sunshine and rain a seed will begin to germinate.

Germinate is the word that describe the process of a seed beginning to grow into a plant. This process can take a day or up to a year depending on the type of seed and weather conditions. In the first part of this process, water softens the seed coat which enables the embryo to grow out of the seed coat.

The first job is to create some roots.

The embryo then grows upwards and creates its first leaf. This leaf is important because the embryo will have used up all of its food from the seed. Once the leaf grows the embryo is called a seedling, the name for a baby plant.

The seedling develops more leaves as the stem grows taller and the roots grow. After many weeks or months the seedling will mature into an adult plant.

The adult plant soon starts to make buds. We call this a budding plant. When the buds open up into flowers we call this the flower plant stage.

At this stage pollination will then occur this is the process for plants where pollen is transferred usually with the help of birds and insects. New seeds will grow with the fruit of the plant. Plants use fruit to help distribute seeds to new locations so the process begins again. This is why we refer to this as the plant life cycle because it goes around and starts again.

Can you complete the stages of the plant cycle. Draw in the plant stage and label it with the corresponding letter.

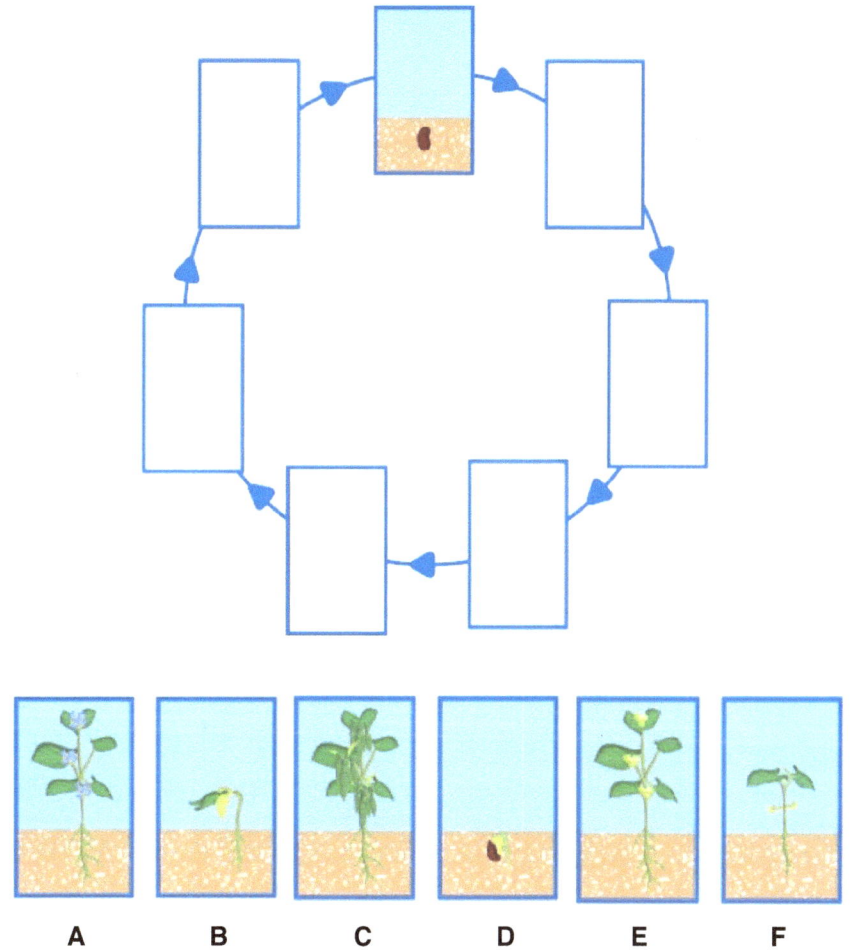

Plants go through the following stages: seed, germination, seedling, budding plant, flowering plant, pollination. Pollination then creates new seeds so the cycle can continue again.

Match the life cycle stage with the purpose.

germination	The stage when the plant produces flowers to attract pollinators.
budding plant	Stage when a seed first starts to grow if the balance of sun and water are correct.
seed	Stage when the seed starts to sprout and grows its first leaves.
seedling	Stage when the plant is fulling grown with buds but no flowers.
pollination	Stage when the plant is a tiny embryo with a built in food supply.
flowering plant	Stage when the plants produces new seeds. Birds, insects and the wind help this to occur.

See if you can match the plant with its description.

annual

A plant that lives for more than two years. Sometimes they survive underground as bulbs or they just stop growing in winter.

biennial

A plant that completes its entire life cycle in one year or less and then dies.

perennial

A plant that takes two years to complete its life cycle. In the first year it produces stems and roots, in the second year it produces flowers and seeds.

Plant Life Cycle Quiz

1. Ferns start as a seed. True or false?

2. The baby plant inside the seed is protect by the _____.
   a. seed coat
   b. root
   c. stem

3. Which part of the seed grows into an adult plant? _____
   a. embryo
   b. seed coat
   c. whole seed

4. The process during which seeds sprout an begin to grow
   is called _____.

5. Once the first leaf grows, the embryo is called a _____.
   a. plant
   b. root
   c. seedling

6. The buds of the plant turn into flowers. True or false?

7. Plants use flowers to distribute seeds. True or false?

8. An annual plant lives more that three years. True or
   false?

States of Water

The Three States of Water.

solid   liquid   Gas

There are three states of water. This means that water can be a solid, a liquid or a gas. You may be familiar with water as a liquid and a solid but might not be familiar with water as a gas. We call this water vapor. Water vapor is an invisible gas and its all around us in the air that we breath.

To demonstrate this, breath some warm air into the palm of your hand you you will notice that its a little bit moist, that's water vapor. If it's a hot sticky day that means that there is a lot of water vapor in the air.

Lets look at how water changes to a solid, to a liquid, to a gas and then back to a liquid. The normal room temperature state of water is as a liquid. When the temperature is below 0^0C or 32^0 F the water freezes and becomes a solid. If the temperature raises above those points, ice melts and becomes a liquid again. Ok, that's pretty straight forward but what about water becoming a gas?

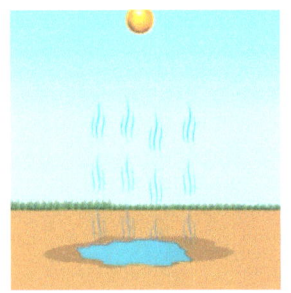

Water becomes a gas whenever its heated. That's the reason a puddle evaporates during a warm day. The puddle gets heat from the sun and then turns into a gas, we call this evaporation. If you heat water on your stove you are speeding up the process of evaporation by giving it more heat so it turns into water vapor quickly. If you heat water to 100^0C or 212^0F you will turn water into gas immediately. This temperature is called water's boiling point.

So how does water vapor turn back into a liquid. Its simple, just reverse the process by cooling the water vapor. If you put a lid on the boiling water and remove the heat the vapor will cool and become droplets. We call this condensation.

Condensation also occurs when you blow on a window an the window steams up. You breath that includes water vapor hits the glass and cools. If you look very closely you will see that the steam on the window is actually many little drops of water.

What are the three states of water.
Label each picture with both the physical state and the description of the water.

gas   liquid   solid   water vapor   water   ice

Put the images in the correct positions. Draw them in the boxes or connect with a line.

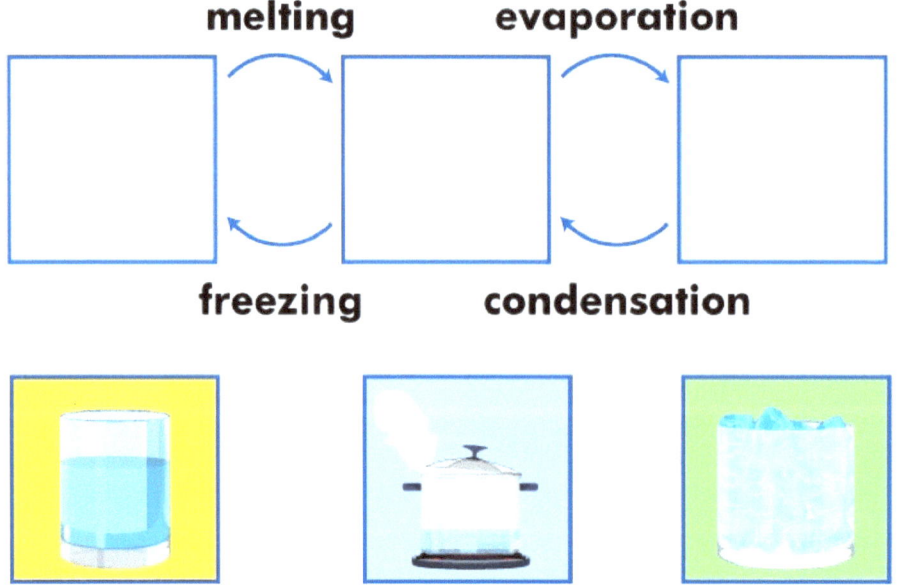

What is the state of water in each example?

The Ice is Melting

That puddle is disappearing on this sunny day.

I blew on the window and it steamed up.

solid to gas

liquid to gas

gas to liquid

gas to solid

solid to liquid

What do we call the change in the state of water?

solid to liquid	
liquid to gas	
gas to liquid	

boiling   **melting**   **condensation**

dripping   **freezing**   **evaporation**

States of Water Quiz

1. The three states of matter are solid, liquid and gas. True or false?

2. Which state does water change into when it is frozen?
    a. gas
    b. liquid
    c. solid
    d. vapor

3. When water changes from a solid to a liquid it is _____?
    a. freezing
    b. melting
    c. evaporation
    d. condensation

4. Evaporation takes place when laundry is dried in a clothes dryer. True or false?

5. Water vapor is an invisible gas. True or false?

6. What state is water when at room temperature? _____.

Changes of Matter

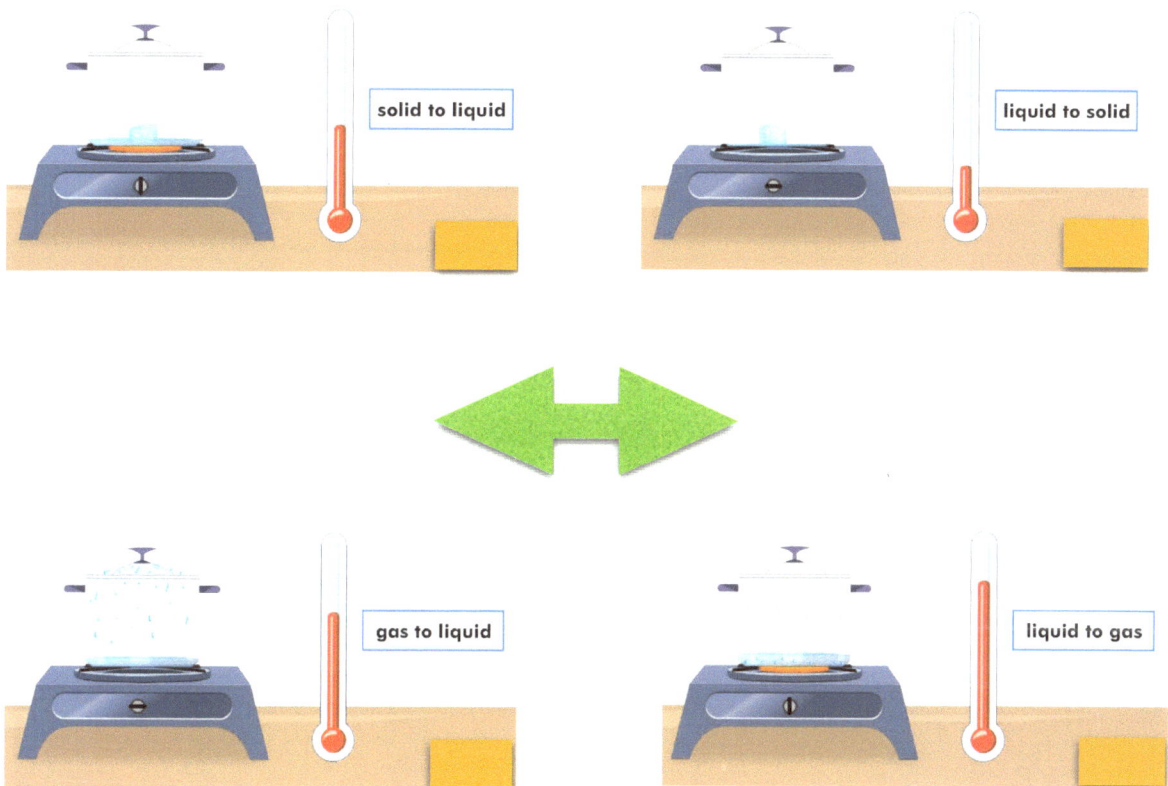

solid to liquid

liquid to solid

gas to liquid

liquid to gas

What change occurs that causes the water to change physical states and then back again?

_____

_____

Photosynthesis is a chemical reaction in which carbon dioxide and water form a chemical reaction in plants to create sugar and oxygen.

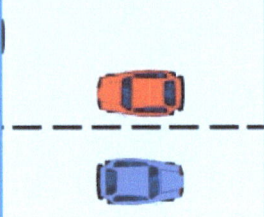

Gasoline burning in our cars is a chemical reaction that produces carbon dioxide, water vapor and a release of energy.

A chemical reaction between baking powered and water creates carbon dioxide gas which causes the cupcake to rise.

Respiration is a chemical reaction in which oxygen and sugar react to react to form carbon dioxide, water vapor and a release of energy.

When wood is burned a chemical reaction occurs between the wood and oxygen, carbon dioxide, water vapor and energy are released.

Iron rusting is a chemical reaction between iron and oxygen in which iron oxide is produced.

Physical Changes vs. Chemical Reaction

A physical change is a change of matter in which the basic particles of a substance remain the same but the physical properties of the item remain the same. For example ice melting is an example of a physical change because liquid water looks different than ice plus it has other differences like its density and hardness. However both ice and liquid water are made up of the same H2O molecules. That's why physics changes are often easy to reverse. For example liquid water can be changed back to ice simply by lowering the temperature.

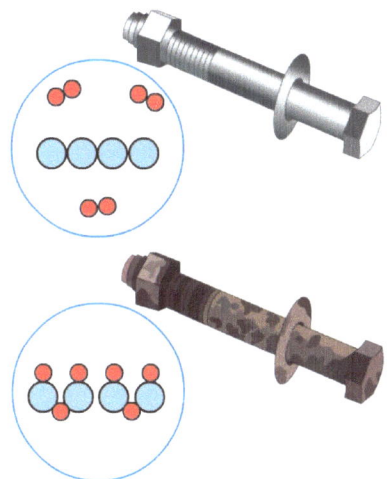

A chemical reaction is a change of matter in which completely new substances are formed. An iron bolt rusting is an example of a chemical reaction. This is because when iron rusts the atoms of iron chemically bond to atoms of oxygen forming a new substance called iron oxide. Iron oxide has complete different properties to iron and can not be easily changed back to iron. Because chemical changes rearrange the atoms of a substance they are not generally reversible.

Many chemical changes occur without us noticing but signs that a chemical change has occurred are a change in color of a substance, a change in odor or the presence of a gas or when light or heat are produced. Remember that when a chemical change occurs matter is never destroyed or created. In ordinary chemical reactions the number atoms of a substance or substances are rearranged an d new substances are formed but the number and types of atoms remains the same even if some of those atoms have moved on to a new location. This is called the law of conservation of matter.

Sort these items by physical or chemical change.

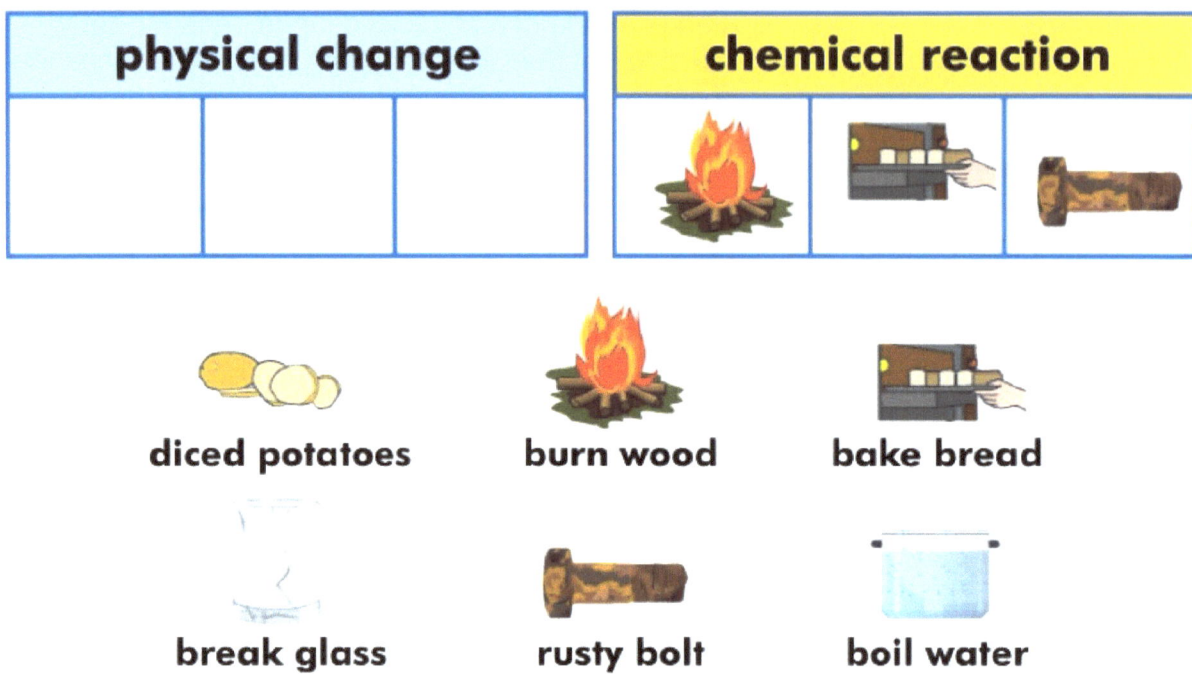

Changes in Matter Quiz

1. All changes of matter are reversible. True or false?

2. Cooking an egg is a reversible change of matter. True or false?

3. A _____ is a change of matter which is not reversible.
   a. physical change
   b. chemical reaction

4. Ice melting is an example of a chemical reaction. True or false?

5. What does the law of conservation of matter say?

_____

_____

_____

Newburyport, MA 01950

1-800-596-3175

OnBoard Academics employs teachers to make lessons for teachers! We create and publish a wide range of aligned lessons in math, science and ELA for use on most EdTech devices including whiteboard, tablets, computers and pdfs for printing.

All of our lessons are aligned to the common core, the Next Generation Science Standards and all state standards.

If you like our products please visit our website for information on individual lessons, teachers licenses, building licenses, district licenses and subscriptions.

Thank you for using OnBoard Academic products.